Welcome to WHALEBONE MANSION

Betcha I can swim faster!

Betcha I can swim deeper!

Welcome to WHALEBONE MANSION

CREATURES THAT LURK AT A WHALE FALL

Laken Slate

Illustrated by
Bindy James

ini Charlesbridge

Did someone dare you to
sneak inside a haunted house?

Be careful. You never know what else may slink through Whalebone Mansion.

This is no house at all. It's a giant sunken skeleton! When a whale dies, its body often sinks. Deep-sea creatures gather at "whale fall" sites to hunt, feast, and hide.

Look, there!
What's that?

Is it smoke from a witch's cauldron?
Is it the reaching arm of a ghost?

It's **the slime** of a snake-like hagfish! Quick, swim away!

The slippery hagfish has no bones. Its skeleton consists entirely of rubbery cartilage. It burrows into its food with rows of hidden, razor-sharp teeth. When it feels threatened, it leaks slime, which mixes with seawater to create an enormous blob. Nearby creatures can get stuck if they're not careful.

At a whale fall,
it's tough to tell
which way is up,
which way is down,
which way is in, and
which way is out.

But the creatures here know their way around.

Creatures such as . . .

. . . the ghoulish goblin shark!
Back up before those big jaws snap shut!

A goblin shark moves slowly, using sensors on its long snout to find prey. Once it has a target, it can extend its jaw away from its face and snatch food using sharp, curly teeth.

Down, down, down!
Away from the vampire squid.

The vampire squid has webbed arms that can squirt clouds of light-up mucus to escape predators. When a vampire squid is frightened, it hides by pulling its lower body completely over its head like a cape. The cape is covered in pointy feelers that look like fangs.

I know it's dark, but don't trust that light.

It may glimmer like a jewel.

It may gleam like sunlight.

But really, it's the . . .

. . . bulb of a hungry anglerfish!

A female anglerfish has a spine that sticks out over its head like a fishing pole. A glowing piece of flesh hangs on the spine to lure prey right into the anglerfish's mouth. Yikes!

Dart through ribs!

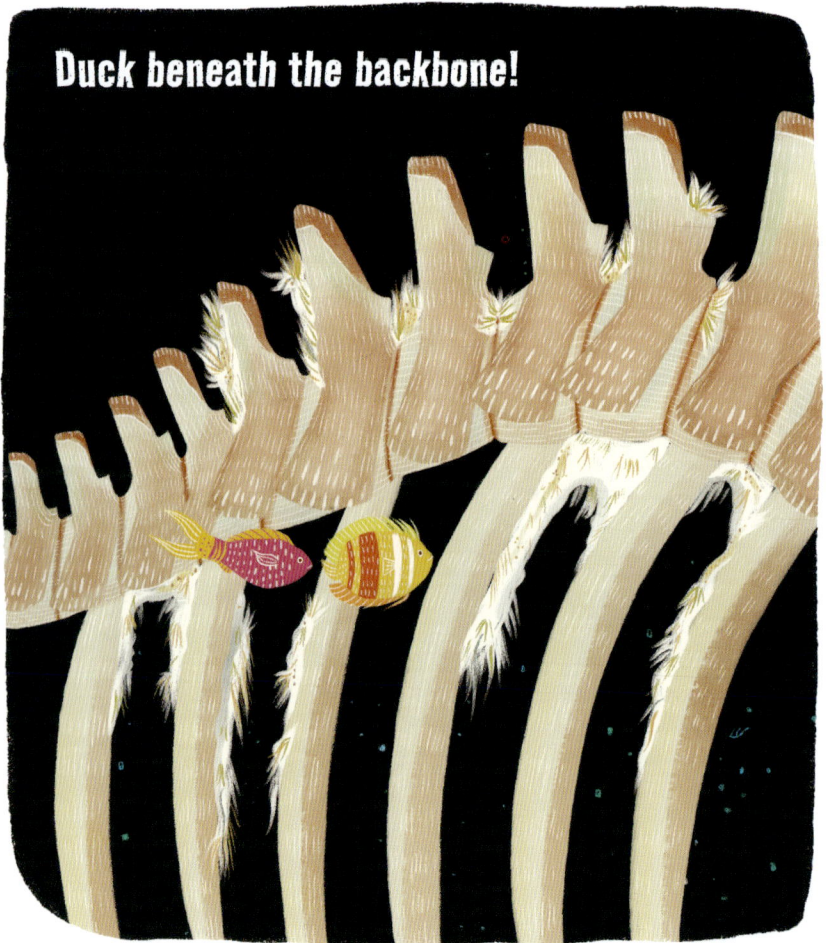

Duck beneath the backbone!

Dodge the dangling tangles
of a dozen octopuses.

Octopuses have sharp
beaks hidden beneath
their sensing arms.
They visit whale falls
to dine on crabs and
the blubber between
the whale's bones.

Whatever you do, don't hide behind that curtain!

A curtain doesn't usually twist.

A curtain doesn't usually wriggle.

It's not a curtain
at all. It's . . .

. . . a bunch of bone-eating zombie worms!

Osedax worms, also called "bone-eating snot flowers," have only ever been found at whale fall sites—no place else. Though they have no mouth or gut, Osedax can group together to consume a whale skeleton hundreds of times their size by dissolving the bones with acid they produce.

past the pinching crustaceans!

Crustaceans such as crabs, shrimp, and lobsters defend themselves against larger beasts on the seafloor. One deep-sea shrimp (*Acanthephyra purpurea*) earned the nickname "fire-breather" when scientists discovered it could spew glow-in-the dark vomit to blind an attacker.

Now you have a whale of a tale
to tell for years to come . . .

as Whalebone Mansion . . .

slowly . . .

Over time, the decaying whale bones release a chemical called sulfur. That draws in rare species, such as clams in the Vesicomyidae family, previously found only near sunken wood and hydrothermal vents (where hot water flows up through cracks in the ocean floor).

. . . disappears.

AUTHOR'S NOTE:

One day I went to the beach, but it was closed. A whale had died, and it was attracting sharks near the shore. I had never considered what happens when one of the largest animals on Earth dies.

I started studying whales, and I learned about whale falls. The photos seemed scary. Giant skeletons underwater! (Of course, humans need a machine to dive deep enough to see a whale fall, and little surface fish can't actually go that deep.)

Hey, I can dive anywhere I like!

Me, too! I'm the best diver!

I read about the many living things that rely on the whale's bones and body for nutrition. A whale fall forms an ecosystem, or community, where some of the rarest beings on the planet interact with one another. Scientists believe whale falls support the evolution of tiny organisms that wouldn't otherwise have access to the nutrients the bones provide.

Most whales live between forty and seventy years. (But bowhead whales can live to be two hundred years old!) Whales spend their lives migrating, hunting, and even singing. When a whale dies, its body brings new life to the ocean's midnight zone.

WHAT IS THE MIDNIGHT ZONE?

The ocean is very deep. Scientists named the layers of the ocean to help humans understand and study each depth. The midnight zone (also called the bathypelagic zone) is the layer between 1,000 meters (about 3,000 feet) and 4,000 meters (about 13,000 feet). A dead whale is considered a "whale fall" when it sinks at least as deep as the midnight zone. Some fall deeper—to the abyssal zone, also called the abyssopelagic zone: 4,000–6,000 meters (13,000–19,700 feet).

In the deep, dark spaces of the ocean, the water is cold enough to preserve a whale's body while deep-sea animals feast and grow. Some of the species that live in the midnight zone have special features like large eyes or bioluminescent (light-up) body parts to help them navigate, eat, and live in the total darkness.

Life has been found at every layer of the ocean, even the Challenger Deep, which is deeper than Mount Everest is tall. (Challenger Deep: 10,935 meters / about 35,900 feet; Mount Everest: 8,850 meters / about 29,000 feet.)

WOULD YOU LIKE TO KNOW MORE ABOUT WHALE FALLS?

Get a poster! https://nmssanctuaries.blob.core.windows.net/sanctuaries-prod/media/mag/5/whale-fall-poster-noaa-onms.pdf. The National Oceanic and Atmospheric Administration (NOAA) released a vivid poster detailing the three confirmed stages of a whale fall, which overlap over time as different creatures feast on and help break down the whale's body:

- mobile-scavenger stage (months to five years after death);
- enrichment-opportunist stage (months to two years after death);
- sulphophilic stage (months to fifty years after death).

There is also a suspected fourth stage that scientists are still figuring out!

Find more books! These are two great books about whale falls available at your local library or bookstore:

- Melissa Stewart and Rob Dunlavey pay poetic tribute to the life and death of whales in their fascinating picture book, *Whale Fall: Exploring an Ocean Floor Ecosystem* (Random House, 2023).
- Jacquie Sewell and Dan Tavis take a comedic angle in their witty, fact-filled picture book, *Whale Fall Café* (Tilbury House, 2021).

Watch a video! Sharon Shattuck, Flora Lichtman, and Sweet Fern Productions staged a delightful paper puppet show called *Life After Whale (On Whale Falls)*. Watch it on the Smithsonian Institution's website: https://ocean.si.edu/ocean-life/marine-mammals/life-after-whale-whale-falls.

DARE YOU TO SWIM INSIDE!

Travel to a real whale fall site.
www.lakenslate.com/resources

SELECTED BIBLIOGRAPHY

Marlow, Jeffrey. "A Whale's Afterlife." *The New Yorker*, February 18, 2019.
www.newyorker.com/science/elements/a-whales-afterlife.

Monterey Bay Aquarium. "From Giants to Gardens: A Fallen Whale's Legacy."
www.montereybayaquarium.org/stories/from-giants-to-gardens.

National Marine Sanctuary Foundation. "Whale Fall 101." January 18, 2021.
www.marinesanctuary.org/blog/whale-fall-101.

National Ocean Service, National Oceanic and Atmospheric Administration. "What Is a Whale Fall?"
www.oceanservice.noaa.gov/facts/whale-fall.html.

To Baron and Audrey, who love spooky stories and the ocean—L. S.

Special thanks to Jeffrey Marlow, Boston University assistant professor of biology and founder and executive director of Ad Astra Academy, an educational outreach project aimed to inspire students around the world. He went out of his way to answer questions I had while writing this book.—L. S.

Printed in China • OPIC
(hc) 10 9 8 7 6 5 4 3 2 1

Charlesbridge • 9 Galen Street, Watertown, MA 02472
www.charlesbridge.com

Illustrations created digitally
Text type set in Road Rage
Edited by Karen Boss
Designed by Jon Simeon and Diane M. Earley
Production supervised by Mira Kennedy

Library of Congress Cataloging-in-Publication Data
Names: Slate, Laken, author. | James, Bindy, illustrator.
Title: Welcome to whalebone mansion / Laken Slate; illustrated by Bindy James.
Description: Watertown, MA: Charlesbridge, [2025] | Includes bibliographical references. | Audience: Ages 3–7 | Audience: Grades K–1 | Summary: "A whale skeleton lies on the bottom of the ocean, and looks a bit like a haunted house as it provides nutrients for deep-water creatures."—Provided by publisher.
Identifiers: LCCN 2024031588 (print) | LCCN 2024031589 (ebook) | ISBN 9781623545789 (hardcover) | ISBN 9781632894496 (ebook)
Subjects: LCSH: Deep-sea ecology—Juvenile literature. | Deep-sea animals—Food—Juvenile literature. | Deep-sea animals—Juvenile literature. | Whales—Juvenile literature.
Classification: LCC QH541.5.D35 S55 2025 (print) | LCC QH541.5.D35 (ebook) | DDC 577.7/9—dc23/eng/20250114
LC record available at https://lccn.loc.gov/2024031588
LC ebook record available at https://lccn.loc.gov/2024031589